Use of this workbook is limited to educational purposes.

This workbook may not be copied, sold, or resold from any person, entity, or organization other than the originating designer, John A. Honeycutt.

All content, images, and thought-leadership are owned by their respective copyright holder and subject to those rights in addition to restrictions imposed by John A. Honeycutt.

Processes associated with design and layout of curriculum is patent-pending by John A. Honeycutt as Honeycutt 21st Century Instructional Design. Design, layout, and content not otherwise owed by other entities is owned by John A. Honeycutt.

HoneycuttScience Work Book

BIOLOGY

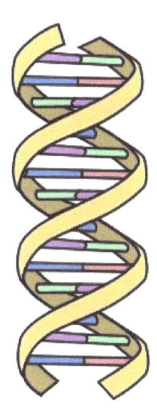

Copyright John A. Honeycutt 2017. All rights reserved.

Contents

11.1	What is Biology?	1
12.1	Scientific Method and Safety	7
13.1	Connections Across Content	13
14.1	Cell Organization	19
15.1	Plant Cells	25
16.1	Prokaryotic and Eukaryotic Cells	31
17.1	Mitosis and Cytokinseis	37
18.1	DNA and Heredity	43
19.1	Genes Genetics and Chromosomes	49
21.1	Organization of Living Things	55
23.1	Interdependence of Living Things	61
24.4	Theory of Evolution	67

11.1 What is Biology?

Summarize main points from each video.

Video Title / topic _____

Video Title / topic _____

Video Title / topic _____

Topic Introduction

Summarize your understanding of each paragraph.

Biology is the study of living organisms. Biology can be divided into many specialized fields. Examples of ways to divide biology into different fields include morphology, physiology, anatomy, behavior, origin, and distribution.

This year, in biology class, we will study a broad set of ideas. There will be several complicated words with most of the topics. For example, the first paragraph on this page lists some unfamiliar terms for most students.

The first paragraph above uses the words "morphology" and "physiology." While these words do not need to be memorized at this time, these words are good examples of how **there will be some new words introduced to you this year.**

The word "anatomy" in the first paragraph is branch of science concerned with the bodily structure of humans, animals, and other living organisms. **Some of the lessons this year will include concepts from anatomy.**

Read/Summarize Text

1. Read the passage.
2. Underline key expressions in each sentence.
3. Re-write each word (or expression) you underlined.
4. Summarize the passage.

Title of Passage.

Definition of a Scientific Law. A scientific law is a statement that describes an observable occurrence in nature that appears to always be true. It is a term used in all of the natural sciences (astronomy, biology, chemistry and physics, to name a few).

The "cell" is an important concept in biology.
Cell is called the fundamental unit of life. In biology, cell theory is the historic scientific theory, now universally accepted, that living organisms are made up of cells. Cells are the basic unit of structure in all organisms and also the basic unit of reproduction.

https://en.wikipedia.org/wiki/Cell_theory

Re-write words you underlined

_____   _____   _____

_____   _____   _____

Using a complete sentence, summarize or rephrase the passage

3

Read Text for Comprehension

Read this article for deeper understanding. No summary is required, although you may want to circle, underline, or mark key ideas and words.

Wikipedia

The Next Generation Science Standards is a multi-state effort to create new education standards that are "rich in content and practice, arranged in a coherent manner across disciplines and grades to provide all students an internationally benchmarked science education.

Overall, the guidelines are intended to help students deeply understand core scientific concepts, to understand the scientific process of developing and testing ideas, and to have a greater ability to evaluate scientific evidence.

The Next Generation Science Standards (NGSS) are based on the "Framework K–12 Science Education" that was created by the National Research Council.

Over 40 states have shown interest in the standards, and as of December 2016, 18 states, along with the District of Columbia (D.C.), have adopted the standards.

Oklahoma Department of Education

The Oklahoma Academic Standards (OAS) for Science focus educators and students on the priority of scientific literacy, so they both appreciate and understand the exceptional nature of science in their everyday lives. This knowledge base and set of skills are essential for our students, so they may be careful consumers of scientific and technical information and have the skills to enter careers in science, engineering, and technology if they so choose. In Oklahoma, the science standards for High School are categorized into three major groups:

- *Physical Science (Physics and Inorganic Chemistry)*
- *Life Science (Biology and Organic Chemistry)*
- *Earth and Space Science (Earth Science and Space Science)*

Honeycutt Science

There are many shared concepts across each of the three major OAS categories. Students reviewing this handout are studying Biology which is part of Life Sciences. Academic standards associated with Life Sciences are addressed in the curriculum you will be studying this year. (Refer also to OAS Science Standards)

https://en.wikipedia.org/wiki/Next_Generation_Science_Standards
http://sde.ok.gov/sde/science

Draw Illustration

Copy and Label the Illustration in the Space Provided

CELL THEORY:

1. All living things are composed of cells.
2. Cells are the basic units of structure and function in living things.
3. All cells are produced from other cells.

http://slideplayer.com/slide/771778/

Draw (Copy) the Illustration Here

Interpret a Graph

Write the title of the graph _____

Circle the type of chart this represents

    Bar Chart   Line Chart   Pie Chart   Other

If applicable,

    What does the X-axis represent _____

    What does the Y-axis imply _____

Summarize what this graph represents or conveys

_____

Reference URL.

Human Heartbeat

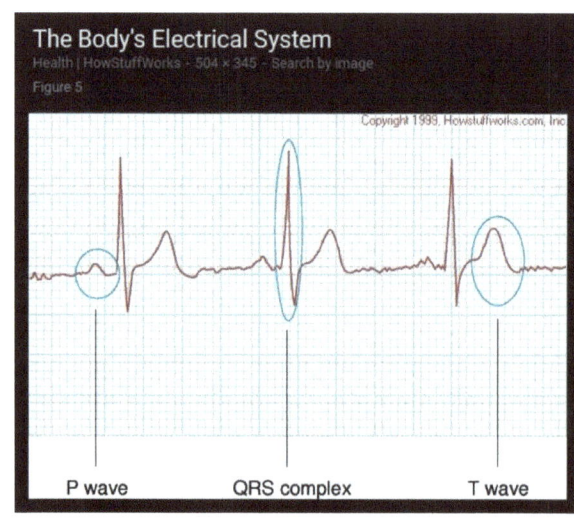

P wave - coincides with the spread of electrical activity over the atria and the beginning of its contraction.

QRS complex - coincides with the spread of electrical activity over the ventricles and the beginning of its contraction.

T wave - coincides with the recovery phase of the ventricles.

http://health.howstuffworks.com/human-body/systems/circulatory/heart4.htm

12.1 Lab Safety and the Scientific Method

Summarize main points from each video.

Video Title / topic

Video Title / topic

Video Title / topic

Topic Introduction

Summarize your understanding of each paragraph.

The scientific method consists of (1) making observations (2) writing down a hypothesis and (3) testing the hypothesis. When new observations do not support the original hypothesis, a new hypothesis is required.

[ ]

Sometimes, hypothesis are formulated before observations are collected; sometimes observations are made before hypothesis are created. Either way, it is important that scientists carefully record their procedures.

[ ]

When a hypothesis is tested in a lab, it is always important to follow safe procedures. In high school labs, students should never "play around" during their work. "Horse-play" and just "joking around" can be dangerous.

[ ]

These tips are always important. Before starting a lab test, study the steps and procedures. Identify the potential risks of each step before starting. Remove or eliminate risks if possible. Always wear appropriate attire and safety gear.

[ ]

Read/Summarize Text

1. Read the passage.
2. Underline key expressions in each sentence.
3. Re-write each word (or expression) you underlined.
4. Summarize the passage.

About scientific hypothesis, observations, theories, and laws.

The scientific method is employed by scientists around the world, but it is not always conducted in the order above. Sometimes, hypothesis are formulated before observations are collected; sometimes observations are made before hypothesis are created. Regardless, it is important that scientists record their procedures carefully, allowing others to reproduce and verify the experimental data and results. After many experiments provide results supporting a hypothesis, the hypothesis becomes a theory. Theories remain theories forever, and are constantly being retested with every experiment and observation.

NOTE: Theories can never become fact or law.

cK12.org

Re-write words you underlined

_____   _____   _____

_____   _____   _____

Using a complete sentence, summarize or rephrase the passage

9

Read Text for Comprehension

Read this article for deeper understanding. No summary is required, although you may want to circle, underline, or mark key ideas and words.

About Experiments

In science, we need to make observations on various phenomena to form and test hypotheses. Some phenomena can be found and studied in nature, but scientists often need to create an experiment.

Experiments are tests under controlled conditions designed to demonstrate something scientists already know or to test something scientists wish to know.

Experiments vary greatly in their goal and scale, but **always rely on repeatable procedure and logical analysis of the results**. The process of designing and performing experiments is a part of the scientific method.

About the Scientific Method

The scientific method is the process used by scientists to acquire new knowledge and improve our understanding of the universe. It involves making observations on the phenomenon being studied, suggesting explanations for the observations, and testing these possible explanations, also called hypotheses, by making new observations. A hypothesis is a scientist's proposed explanation of a phenomenon which still must be tested.

Contrast of Scientific Theories and Laws

In science, a law is a mathematical relationship that exists between observations under a given set of conditions.

There is a fundamental difference between observations of the physical world and explanations of the nature of the physical world. Hypotheses and theories are explanations, whereas laws and measurements are observational.

Explanations	Observational
Theories & Hypothesis	Scientific Law

cK12.org

Draw Illustration

Copy and Label the Illustration in the Space Provided

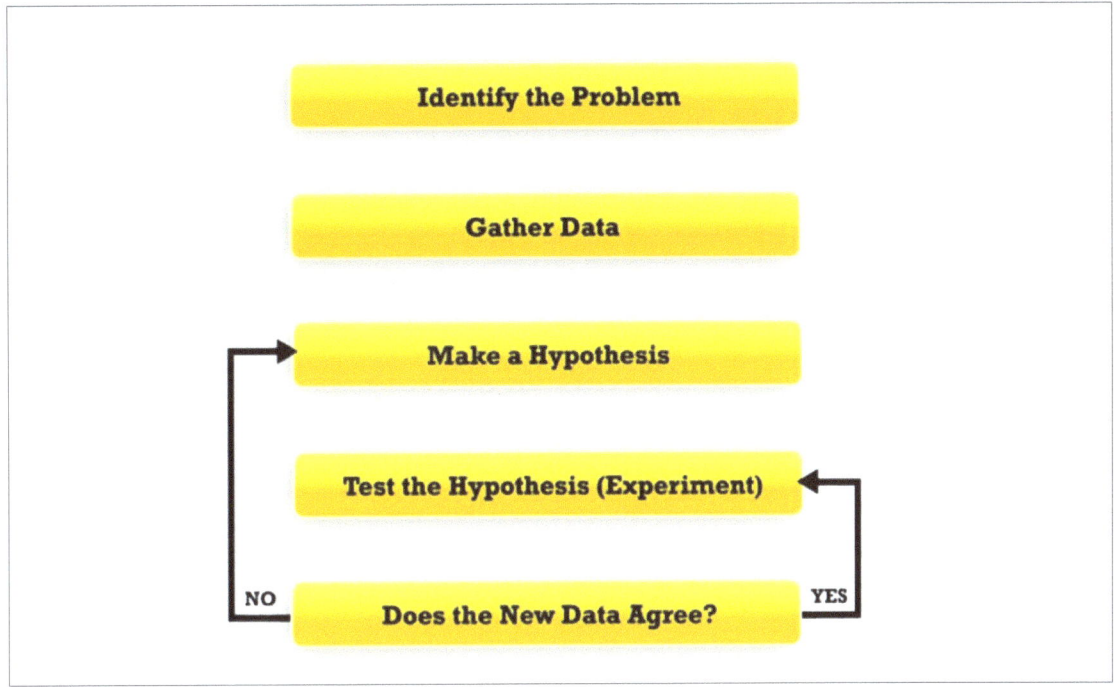

cK12.org

Draw (Copy) the Illustration Here

Interpret a Graph

Write the title of the graph _____

Circle the type of chart this represents
       Bar Chart   Line Chart   Pie Chart   Other

If applicable,
       What does the X-axis represent _____

       What does the Y-axis imply _____

Summarize what this graph represents or conveys

_____

http://cen.acs.org

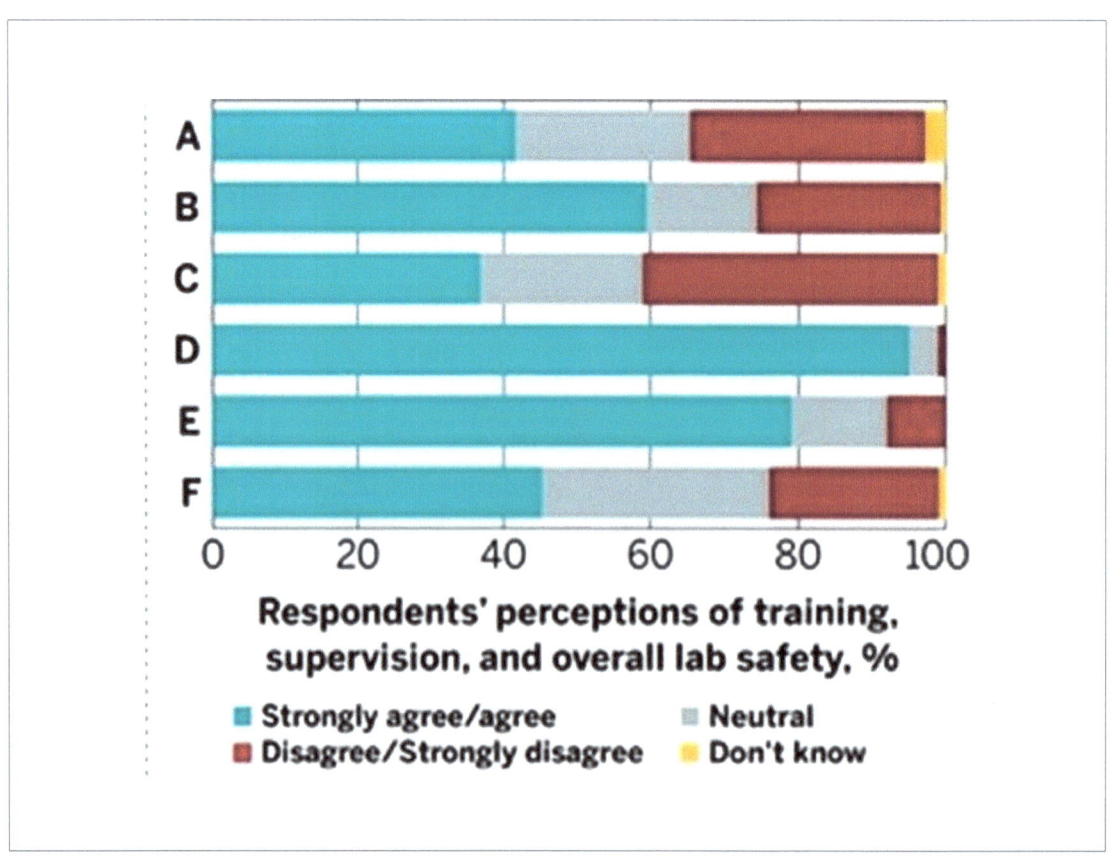

13.1 Connections Across Content

Summarize main points from each video.

Video Title / topic

Video Title / topic

Video Title / topic

Topic Introduction

Summarize your understanding of each paragraph.

Science is a logical activity. Science identifies, builds and organizes knowledge. Science knowledge can be tested. Scientists test ideas that other scientists have identified through observation. Through science, people can explain and predict things about the universe.

Modern science is broadly divided into the natural sciences and the social sciences. This class (biology) along with physical science, earth science, and chemistry are all natural sciences. Natural science studies and deals with the material world.

While this class is NOT a social science class, you may be interested to learn that social sciences deal with the study of people and societies. Psychology, sociology, anthropology, and history are among the many social sciences. (You do not need to recall this paragraph).

Natural science can be divided into two main branches: life science and physical science. You are studying biology which is a "life science" – the study of living things. Other students in the school are studying physical science (the study of non-living things.)

Read/Summarize Text

1. **Read the passage.**
2. **Underline key expressions in each sentence.**
3. **Re-write each word (or expression) you underlined.**
4. **Summarize the passage.**

High School Science (at Keota)

> Biology is taught as a "life science" – whereas the other three science classes at our school are as a "physical science."
>
> Four natural science subjects are presently taught at our school. Biology, physical science, earth/space science, and chemistry are each considered natural sciences. But, biology is the only "life science" course presently taught in our school. The other three are taught as "physical sciences."
>
> Still, there are some connections between biology, earth science, and chemistry. The earth, obviously supports life. And life is made up of atoms and molecules (chemistry).

HoneycuttScience.com

Re-write words you underlined

_____   _____   _____

_____   _____   _____

Using a complete sentence, summarize or rephrase the passage

15

Read Text for Comprehension

Read this article for deeper understanding. No summary is required, although you may want to circle, underline, or mark key ideas and words.

More About Biology ...

Biology - This field encompasses a set of disciplines that examines phenomena related to living organisms. The scale of study can range from sub-component biophysics up to complex ecologies. Biology is concerned with the characteristics, classification and behaviors of organisms, as well as how species were formed and their interactions with each other and the environment.

Biological Fields of Study

The biological fields of botany, zoology, and medicine date back to early periods of civilization, while microbiology was introduced in the 17th century with the invention of the microscope. However, it was not until the 19th century that biology became a unified science. Once scientists discovered commonalities between all living things, it was decided they were best studied as a whole.

Key Developments in Biology

Some key developments in biology were the discovery of genetics; Darwin's theory of evolution through natural selection; the germ theory of disease and the application of the techniques of chemistry and physics at the level of the cell or organic molecule.

Sub-disciplines of Biology

Modern biology is divided into sub-disciplines by the type of organism and by the scale being studied. Molecular biology is the study of the fundamental chemistry of life, while cellular biology is the examination of the cell; the basic building block of all life. At a higher level, anatomy and physiology looks at the internal structures, and their functions, of an organism, while ecology looks at how various organisms interrelate.

https://en.wikipedia.org/wiki/Natural_science

Draw Illustration

Copy and Label the Illustration in the Space Provided

Illustration

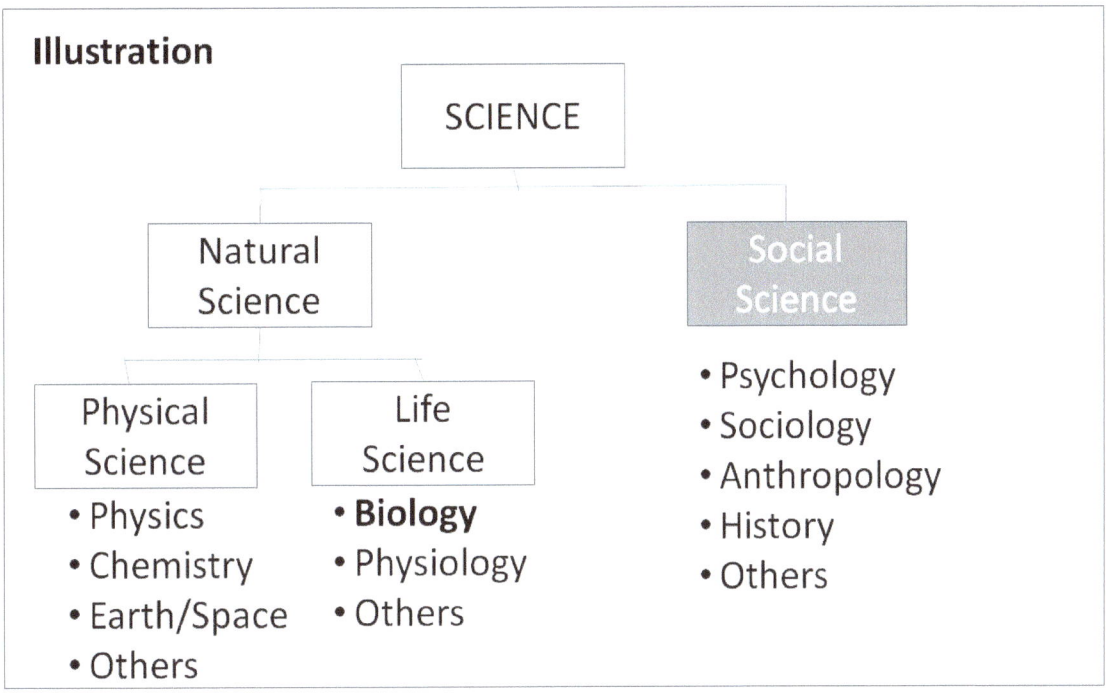

www.HoneycuttScience.com

Draw (Copy) the Illustration Here

Interpret a Graph

Write the title of the graph _____

Circle the type of chart this represents

    Bar Chart   Line Chart   Pie Chart   Other

If applicable,

    What does the X-axis represent _____

    What does the Y-axis imply _____

Summarize what this graph represents or conveys

_____

http://www.statcan.gc.ca

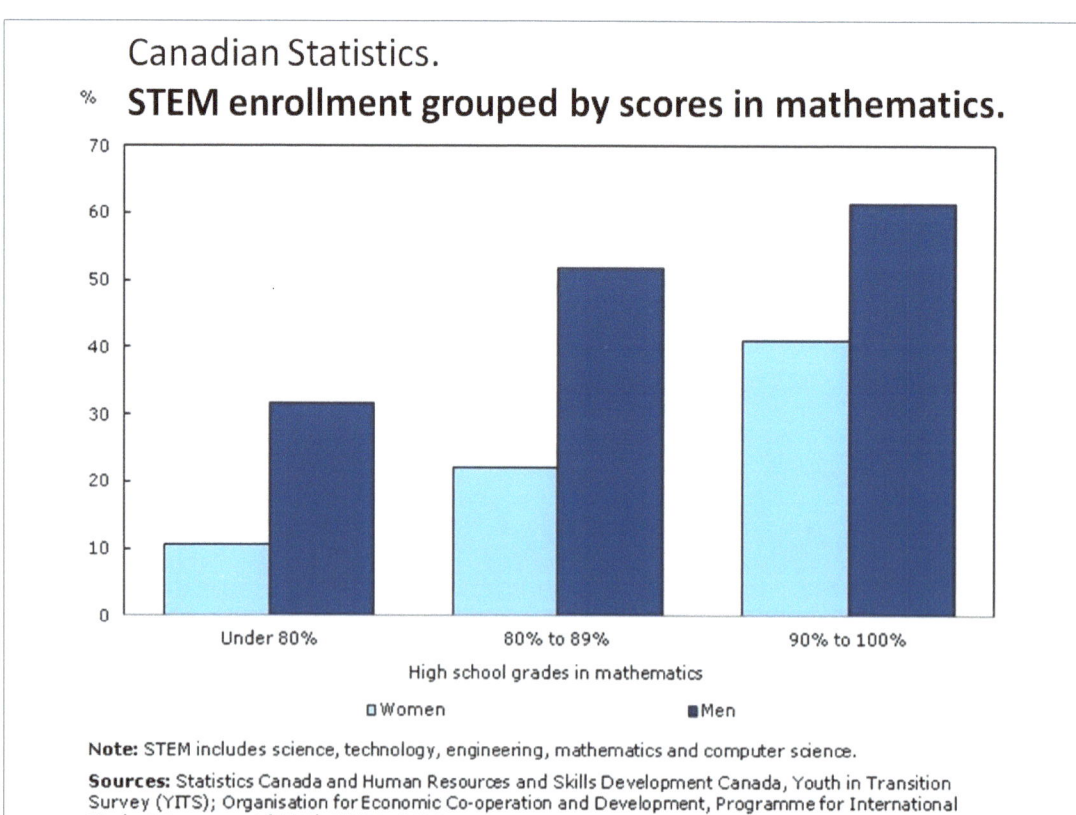

14.1 Cell Organization

Summarize main points from each video.

Video Title / topic _____

Video Title / topic _____

Video Title / topic _____

Topic Introduction

Summarize your understanding of each paragraph.

The word "cell" comes from a Latin word. Originally a Latin word was used. The Latin word was "cella." The Latin word "cella" means "small room." In English speaking countries, we translate the Latin word to English. We call them cells.

In the year 1665, a man named Robert Hooke discovered cells in cork. Later, he found them in living plant tissue. He found the cells by using a compound microscope. Robert Hooke used the Latin word cella to describe these "small rooms."

Cells are the basic unit of all known living organisms. Cells are the basic unit of all known life. Cells are the basic structural, functional, and biological unit of life. A cell is the smallest unit of life that can independently replicate.

Cell biology is a branch of biology. Cell biology studies the structures and functions of cells. Cell biology focuses on the idea that the cell is the basic unit of life. Cell biology explains the structure, components and organization of cells. It studies how cells interact.

Read/Summarize Text

1. Read the passage.
2. Underline key expressions in each sentence.
3. Re-write each word (or expression) you underlined.
4. Summarize the passage.

The First Cell ...

> There are several theories about the origin of small molecules that led to life on the early Earth. They may have been carried to Earth on meteorites, created at deep-sea vents, or synthesized by lightning in a reducing atmosphere.
>
> There is not much experimental data defining what the first self-replicating forms were.
>
> RNA is thought to be the earliest self-replicating molecule, as it is capable of both storing genetic information and catalyzing chemical reactions. But some other entity with the potential to self-replicate could have preceded RNA, such as clay or peptide nucleic acid.

https://en.wikipedia.org/wiki/Cell_(biology)

Re-write words you underlined

_____ _____ _____

_____ _____ _____

Using a complete sentence, summarize or rephrase the passage

Read Text for Comprehension

Read this article for deeper understanding. No summary is required, although you may want to circle, underline, or mark key ideas and words.

You need to remember two words. You need to know how to spell both of these words. These words are probably new to you. You will learn more about both of these words later. For the moment though – make sure you know how to pronounce AND spell both of these words:

Words	Pronunciation			
Prokaryotic.	PRO	kari	ah	tik
Eukaryotic.	YOU	kari	ah	tik

All cells, whether prokaryotic or eukaryotic, have a membrane that envelops the cell, and regulates what moves in and out of the cell.

NOTE: The expression "selectively permeable" is used to describe this "regulation." The membrane that surrounds the cell is selectively permeable.

Inside the membrane, the cytoplasm takes up most of the cell's volume. Many cells possess DNA. DNA is the hereditary material of genes.

Another important abbreviation to memorize is RNA. RNA contains information necessary to build various proteins such as enzymes.

Become familiar with these words, expressions, and abbreviations:

Prokaryotic

Eukaryotic

Membrane

Cytoplasm

DNA

RNA

Proteins

Enzymes

https://en.wikipedia.org/wiki/Cell_(biology)

Draw Illustration

Copy and Label the Illustration in the Space Provided

Illustration

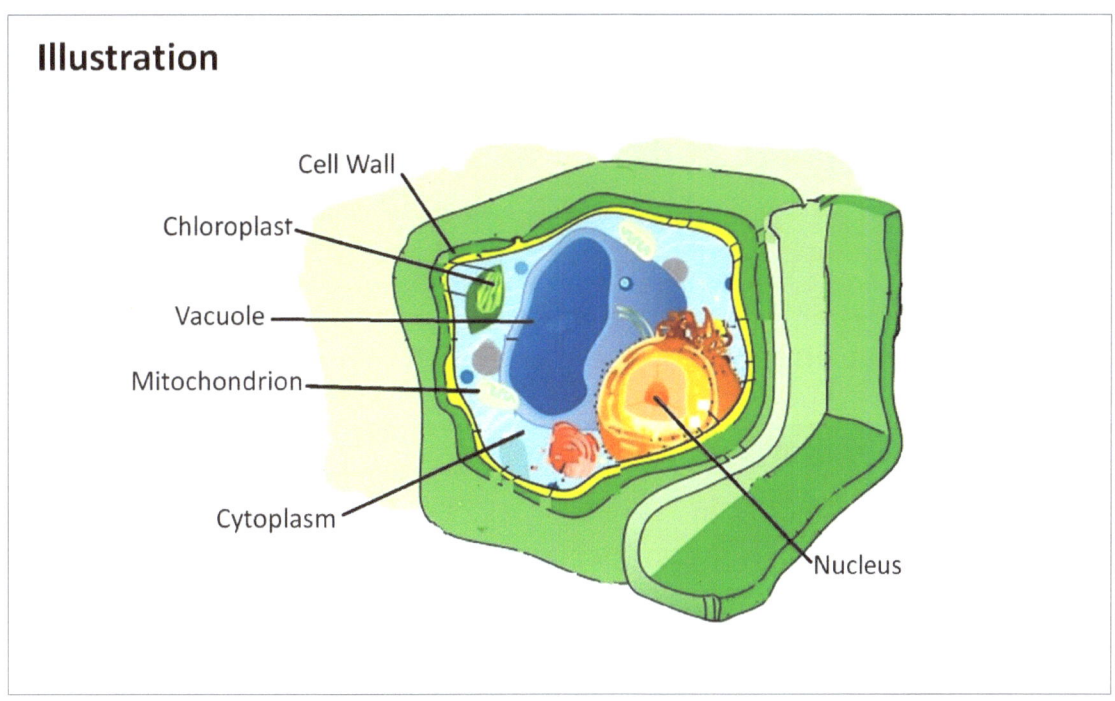

Adapted from https://en.wikipedia.org/wiki/Cell_(biology)#/media/File:Plant_cell_structure-en.svg

Interpret a Graph

Write the title of the graph _____

Circle the type of chart this represents

     Bar Chart   Line Chart   Pie Chart   Other

If applicable,

    What does the X-axis represent _____

    What does the Y-axis imply _____

Summarize what this graph represents or conveys

_____

https://www.sciencedaily.com/releases/2011/08/110823180459.htm

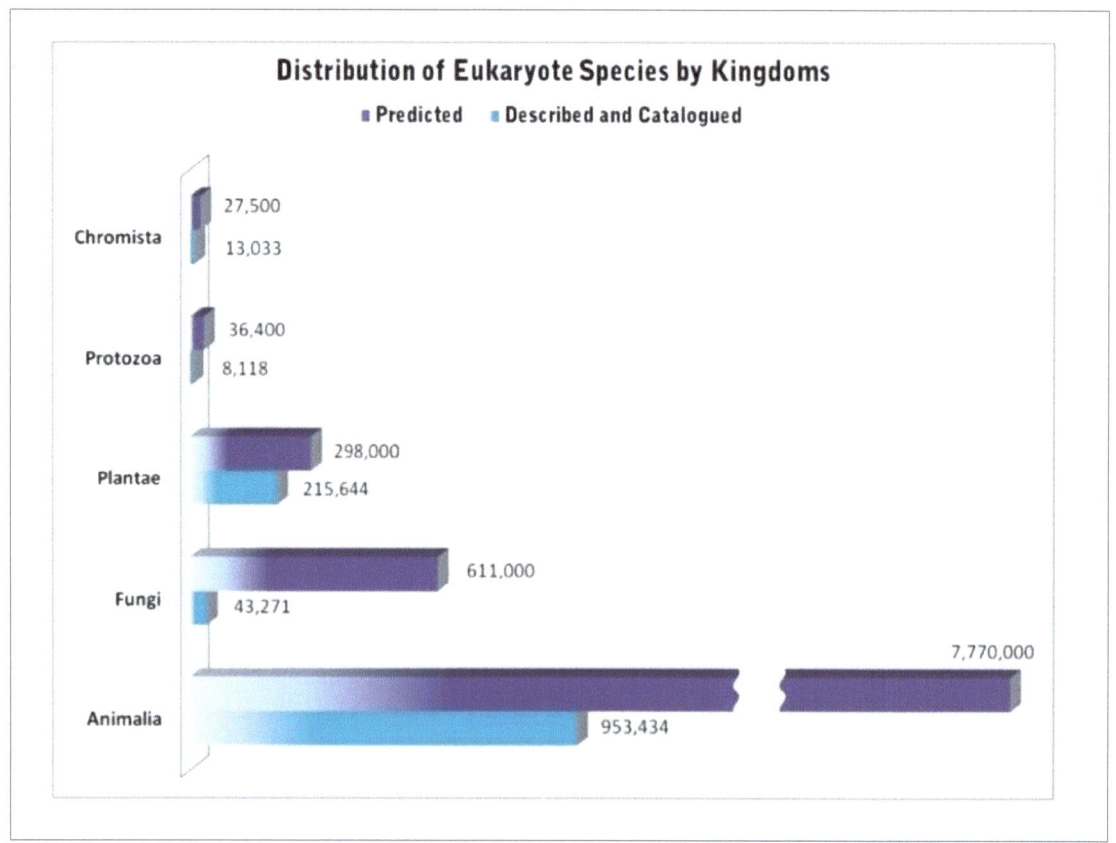

15.1 Plant Cells

Summarize main points from each video.

Video Title / topic

Video Title / topic

Video Title / topic

Topic Introduction

Summarize your understanding of each paragraph.

Genetically modified food controversies are disputes over the use of foods and other goods derived from genetically modified crops. Genetically modified food is sometimes used instead of conventional crops.

The acronym "GMO" stands for "genetically modified organism." Disputes over GMOs involve consumers, farmers, biotechnology companies, governmental regulators, non-governmental organizations, and scientists.

Specific concerns from some groups regarding GMOs the include mixing of genetically modified and non-genetically modified products in the food supply. Some are concerned with potential effects of GMOs on the environment.

As of this date (Sept. 2017), no authoritative reports of ill effects have been documented in the human population from genetically modified food. There is a scientific consensus that food derived to-date from GM crops poses no greater risk to health than conventional food.

https://en.wikipedia.org/wiki/Genetically_modified_food_controversies

Read/Summarize Text

1. Read the passage.
2. Underline key expressions in each sentence.
3. Re-write each word (or expression) you underlined.
4. Summarize the passage.

GMO information

> Consumer concerns about food quality first became prominent long before the advent of GM foods in the 1990s. Upton Sinclair's novel The Jungle led to the 1906 Pure Food and Drug Act, the first major US legislation on the subject.
>
> This began an enduring concern over the purity and later "naturalness" of food that evolved from a focus on sanitation to include added ingredients such as preservatives and flavors and sweeteners, residues such as pesticides, the rise of organic food as a category and finally to concerns over GM food. The public came to see the latter as "unnatural."

https://en.wikipedia.org/wiki/Genetically_modified_food_controversies

Re-write words you underlined

_____   _____   _____

_____   _____   _____

Using a complete sentence, summarize or rephrase the passage

27

Read Text for Comprehension

Read this article for deeper understanding. Inserted genes can be classified into three groups based on their use. Read below, for more ...

Genes That Protect a Crop.

The major use of plant genetic engineering has been to make crops easier to grow by decreasing the impact of pests. Insect resistance has been achieved by transforming a crop using a Bt gene. Bt genes were isolated from Bacillus thuringiensis, a common soil bacterium. They code for proteins that severely disrupt the digestive system of insects. Thus an insect eating the leaf of a plant expressing a Bt gene stops eating and dies of starvation. There are many Bt genes, each of which targets a particular group of insects. Some Bt genes, for example, target caterpillars. Others target beetles.

Genetic engineering also has been used in the battle against weeds. Bacterial genes allow crops to either degrade herbicides or be resistant to them. The herbicides that are used are generally very effective, killing most plants. They are considered environmentally benign, degrading rapidly in the soil and having little impact on humans or other organisms. Thus a whole field of transgenic crops can be sprayed with broad-spectrum herbicides, killing all plants except the crops. Corn, soybeans, canola, and cotton that have been engineered to withstand either insects or herbicides, or both, are widely planted in some countries, including the United States. In addition, other crops, including potatoes, tomatoes, tropical fruits, and melons, have been engineered for resistance to viral diseases.

Genes That Improve Crop Quality.

An emerging major use of genetic engineering for crops is to alter the quality of the crop. Fresh fruits and vegetables begin to deteriorate immediately after being harvested. Delaying or preventing this deterioration not only preserves a produce's flavor, and appearance, but maintains the nutritional value of the produce. Genes that change the hormonal status of the harvested crops are the major targets for genetic engineering toward longer shelf-life.

Genes That Introduce New Traits.

One approach to improving the economic value of crops is to give them traits that are completely new for that plant. Some crops, including potatoes, tomatoes, and bananas, have been engineered with genes from pathogenic organisms. This is done to make animals, including humans, that eat the crops immune to the diseases caused by the pathogens.

http://medicine.jrank.org/pages/2902/Transgenic-Plants.html

Draw Illustration

Copy and Label the Illustration in the Space Provided

Illustration

- Gene Silencing
- Applied Biosystems
- Gene Sequencing
- Gene Synthesis
- Human Genome Sequence
- GMO Crops
- Genetically Altered Foods

Draw (Copy) the Illustration Here

http://medicine.jrank.org/pages/2902/Transgenic-Plants.html

Interpret a Graph

Write the title of the graph _____

Circle the type of chart this represents

        Bar Chart   Line Chart   Pie Chart   Other

If applicable,

    What does the X-axis represent _____

    What does the Y-axis imply _____

Summarize what this graph represents or conveys

_____

http://time.com/3840073/gmo-food-charts/

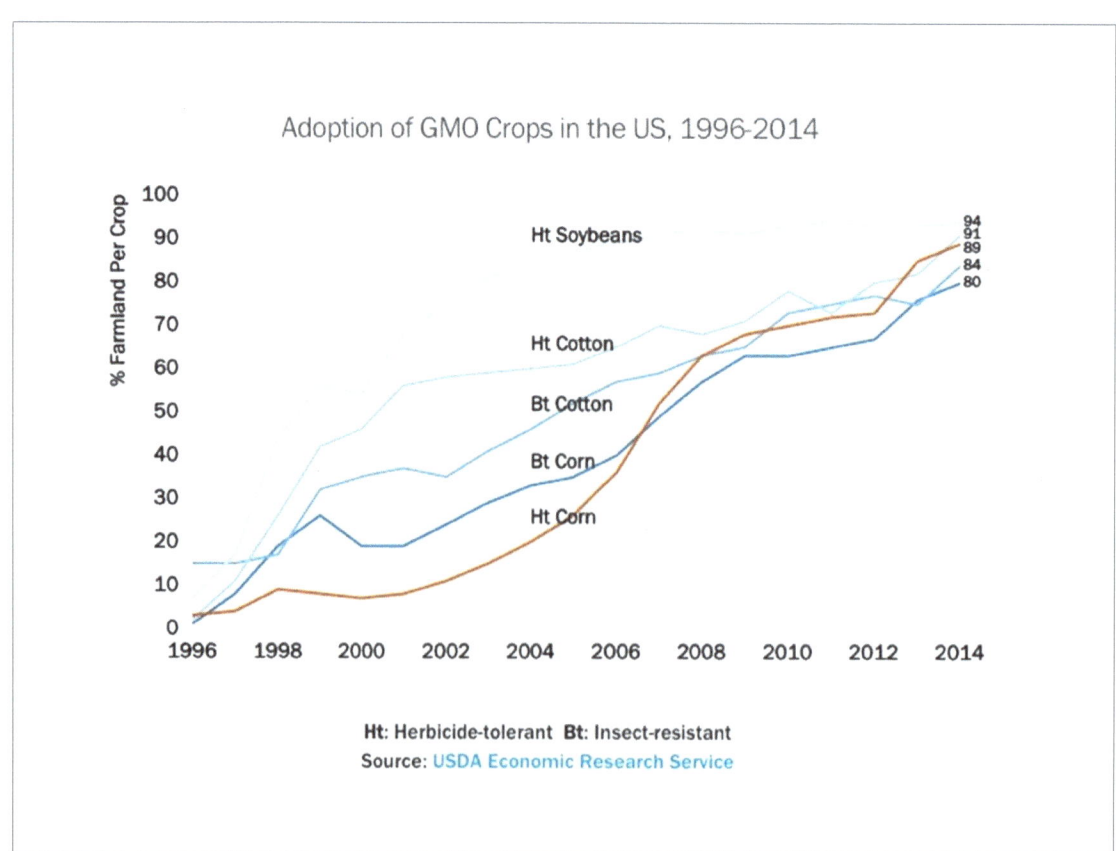

16.1 Prokaryotic and Eukaryotic Cells

Summarize main points from each video.

Video Title / topic

Video Title / topic

Video Title / topic

Topic Introduction

Summarize your understanding of each paragraph.

Prokaryotes are unicellular organisms that lack organelles or other internal membrane-bound structures. They do not have a nucleus. Instead, they have a single chromosome.

[ ]

One hypothesis is that eukaryotic cells evolved from a symbiotic association of prokaryotes—endosymbiosis. This is well supported by studies of mitochondria and chloroplasts, which are thought to have evolved from bacteria living in large cells.

[ ]

Prokaryotes are found in the domains of Bacteria and Archaea. Eukaryotes make up the remaining domain. Prokaryotes tend to be much smaller in size than eukaryotic cells. Prokaryotes have no membrane-bound organelles such as a nucleus.

[ ]

Autotrophic prokaryotes make organic molecules from carbon dioxide. In contrast, heterotrophic prokaryotes obtain carbon from organic compounds.

[ ]

Read/Summarize Text

1. Read the passage.
2. Underline key expressions in each sentence.
3. Re-write each word (or expression) you underlined.
4. Summarize the passage.

From prokaryotes to eukaryotes.

> Mitochondria and chloroplasts have striking similarities to bacteria cells. They have their own DNA, which is separate from the DNA found in the nucleus of the cell. And both organelles use their DNA to produce many proteins and enzymes required for their function.
>
> Living things have evolved into three large clusters of closely related organisms, called "domains": Archaea, Bacteria, and Eukaryota. Archaea and Bacteria are small, relatively simple cells surrounded by a membrane and a cell wall, with a circular strand of DNA containing their genes.

http://learn.genetics.utah.edu/content/cells/organelles/

Re-write words you underlined

_____ _____ _____

_____ _____ _____

Using a complete sentence, summarize or rephrase the passage

Read Text for Comprehension

Read this article for deeper understanding. No summary is required, although you may want to circle, underline, or mark key ideas and words.

Take a moment and look at yourself. How many organisms do you see? Your first thought might be that there's just one: yourself. However, if you were to look closer, at the surface of your skin or inside your digestive tract, you would see that there are actually many organisms living there. That's right! You are home to around 100 trillion bacterial cells, which outnumber your own human cells by about 10 to one1

This means that your body is actually an ecosystem. It also means that you—for some definition of the word you—actually consist of both of the major types of cells: prokaryotic and eukaryotic.

All cells fall into one of these two broad categories. Only the single-celled organisms of the domains Bacteria and Archaea are classified as prokaryotes—pro means before and kary means nucleus. Animals, plants, fungi, and protists are all eukaryotes—eu means true—and are made up of eukaryotic cells. Often, though—as in the case of we humans—there are some prokaryotic friends hanging around.

Components of prokaryotic cells

There are some key ingredients that a cell needs in order to be a cell, regardless of whether it is prokaryotic or eukaryotic. All cells share four key components:

- The plasma membrane is an outer covering that separates the cell's interior from its surrounding environment.
- Cytoplasm consists of the jelly-like cytosol inside the cell, plus the cellular structures suspended in it. In eukaryotes, cytoplasm specifically means the region outside the nucleus but inside the plasma membrane.
- DNA is the genetic material of the cell.
- Ribosomes are molecular machines that synthesize proteins.

Despite these similarities, prokaryotes and eukaryotes differ in a number of important ways. A prokaryote is a simple, single-celled organism that lacks a nucleus and membrane-bound organelles.

https://www.khanacademy.org/science/biology/structure-of-a-cell/prokaryotic-and-eukaryotic-cells/a/prokaryotic-cells

Draw Illustration

Copy and Label the Illustration in the Space Provided

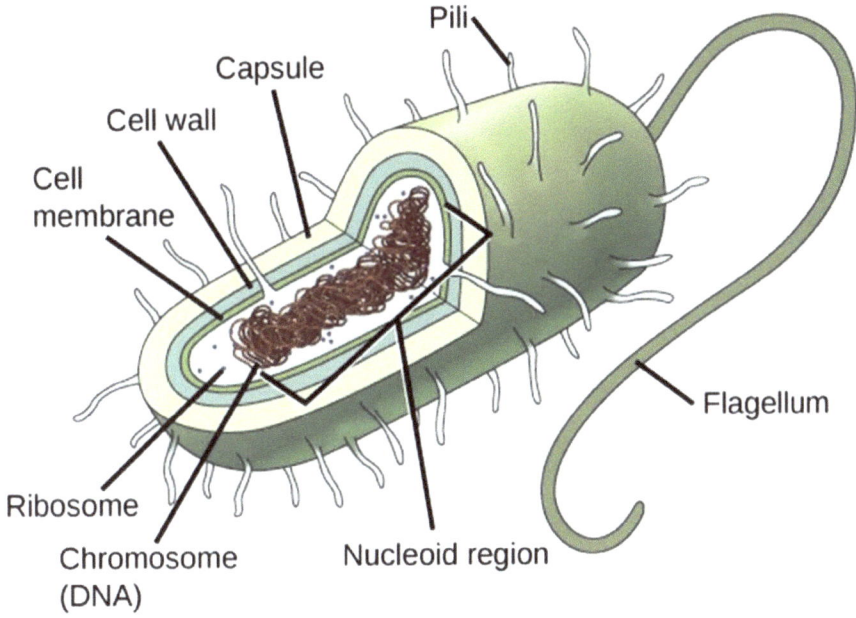

NOTE: This figure shows the generalized structure of a prokaryotic cell. All prokaryotes have chromosomal DNA localized in a nucleoid, ribosomes, a cell membrane, and a cell wall. The other structures shown are present in some, but not all, bacteria.

http://cnx.org/contents/GFy_h8cu@9.87:pOpVdIwp@11/Prokaryotic-Cells

Draw (Copy) the Illustration Here

Interpret a Graph

Write the title of the graph _____

Circle the type of chart this represents
        Bar Chart   Line Chart   Pie Chart   Other

If applicable,
    What does the X-axis represent _____

    What does the Y-axis imply _____

Summarize what this graph represents or conveys

_____

https://www.ck12.org/biology/Prokaryotic-and-Eukaryotic-Cells/lesson/Two-Types-of-Cells-Advanced-BIO-ADV/

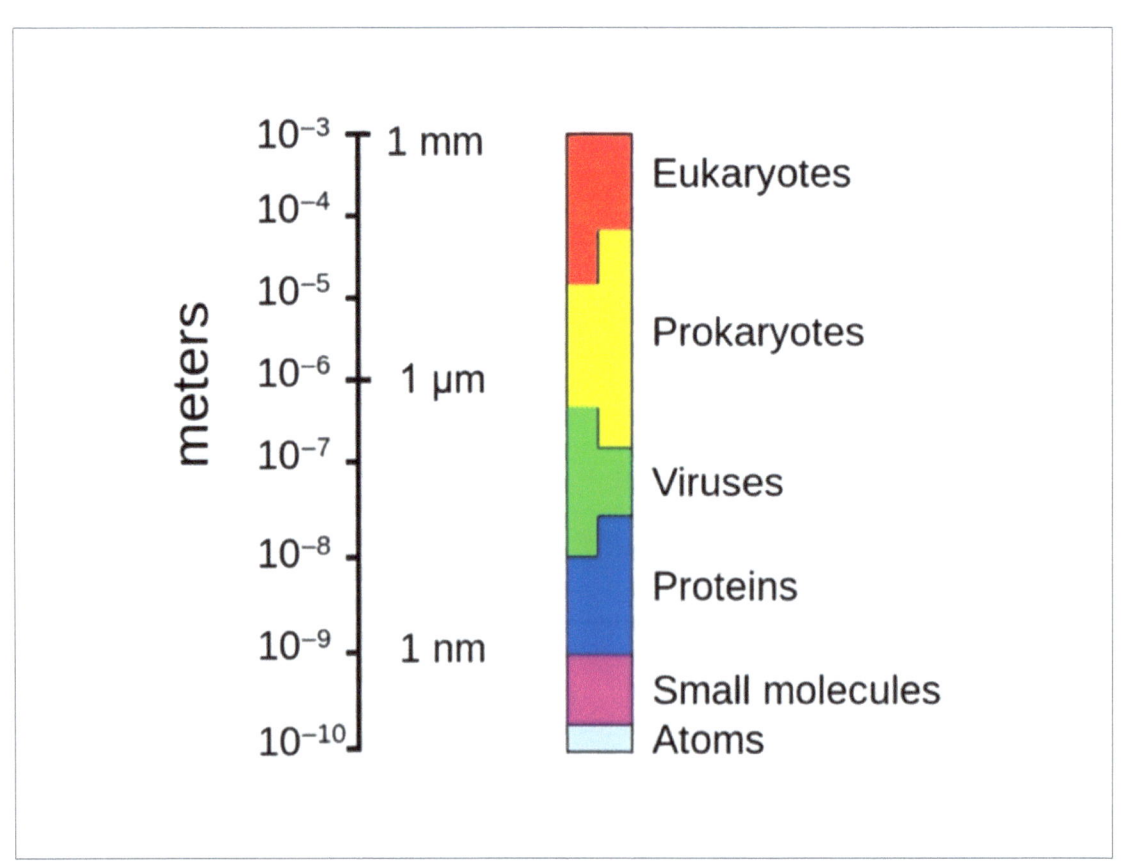

17.1 Mitosis and Cytokinseis

Summarize main points from each video.

Video Title / topic _____

Video Title / topic _____

Video Title / topic _____

Topic Introduction

Summarize your understanding of each paragraph.

During mitosis, the first and longest phase of mitosis is prophase. During prophase, chromatin condenses into chromosomes, and the nuclear envelope, or membrane, breaks down.

>

Next, during metaphase, spindle fibers attach to the centromere of each pair of sister chromatids. The sister chromatids line up at the equator, or center, of the cell. Spindle fibers move sister chromatids to separate and go to different daughter cells when the cell divides.

>

Then, during anaphase, sister chromatids separate and the centromeres divide. The sister chromatids are pulled apart by the shortening of the spindle fibers. One way to imagine this is that this is similar to reeling in a fish by shortening the fishing line.

>

Finally, during telophase, the chromosomes begin to uncoil and form chromatin. This prepares the genetic material for directing the metabolic activities of the new cells. The spindle also breaks down, and new nuclear membranes (nuclear envelope) form.

>

https://www.ck12.org/biology/mitosis

Read/Summarize Text

1. Read the passage.
2. Underline key expressions in each sentence.
3. Re-write each word (or expression) you underlined.
4. Summarize the passage.

Title of Passage.

> Mitosis actually occurs in four phases. The phases are called prophase, metaphase, anaphase, and telophase. They are described in the paragraphs above and illustrated in the diagrams following this page.
>
> Mitosis is the phase of the eukaryotic cell cycle that occurs between DNA replication and the formation of two daughter cells.
>
> There are a few differences between plant and animal cells during mitosis. Muck of the process is the same, however.

https://www.ck12.org/biology/mitosis/lesson/Mitosis-and-Cytokinesis-BIO/

Re-write words you underlined

_____   _____   _____

_____   _____   _____

Using a complete sentence, summarize or rephrase the passage

Read Text for Comprehension

Read this article for deeper understanding. No summary is required, although you may want to circle, underline, or mark key ideas and words.

In cell biology, mitosis is a part of the cell cycle when replicated chromosomes are separated into two new nuclei. In general, mitosis (division of the nucleus) is preceded by the S stage of interphase (during which the DNA is replicated) and is often accompanied or followed by cytokinesis, which divides the cytoplasm, organelles and cell membrane into two new cells containing roughly equal shares of these cellular components. Mitosis and cytokinesis together define the mitotic phase of an animal cell cycle—the division of the mother cell into two daughter cells genetically identical to each other.

The process of mitosis is divided into stages corresponding to the completion of one set of activities and the start of the next. These stages are prophase, prometaphase, metaphase, anaphase, and telophase. During mitosis, the chromosomes, which have already duplicated, condense and attach to spindle fibers that pull one copy of each chromosome to opposite sides of the cell. The result is two genetically identical daughter nuclei. The rest of the cell may then continue to divide by cytokinesis to produce two daughter cells.

Mitosis occurs only in eukaryotic cells. Prokaryotic cells, which lack a nucleus, divide by a different process called binary fission. Mitosis varies between organisms. For example, animal cells undergo an "open" mitosis, where the nuclear envelope breaks down before the chromosomes separate, whereas fungi undergo a "closed" mitosis, where chromosomes divide within an intact cell nucleus. Most animal cells undergo a shape change, known as mitotic cell rounding, to adopt a near spherical morphology at the start of mitosis. Most human cells are produced by mitotic cell division. Important exceptions include the gametes – sperm and egg cells – which are produced by meiosis.

https://en.wikipedia.org/wiki/Mitosis

Mitosis and Cytokinesis

Copy and Label the Illustration in the Space Provided

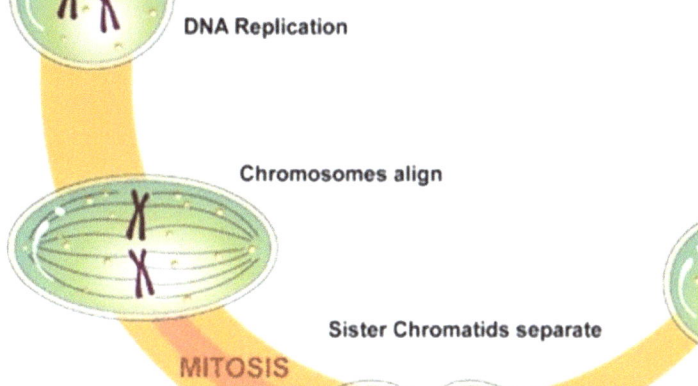

DNA Replication

Chromosomes align

Sister Chromatids separate

MITOSIS

2 Diploid Cells

https://www.ck12.org/biology/mitosis/lesson/Mitosis-and-Cytokinesis-BIO/

Draw (Copy) the Illustration Here

Mitosis and Cytokinesis

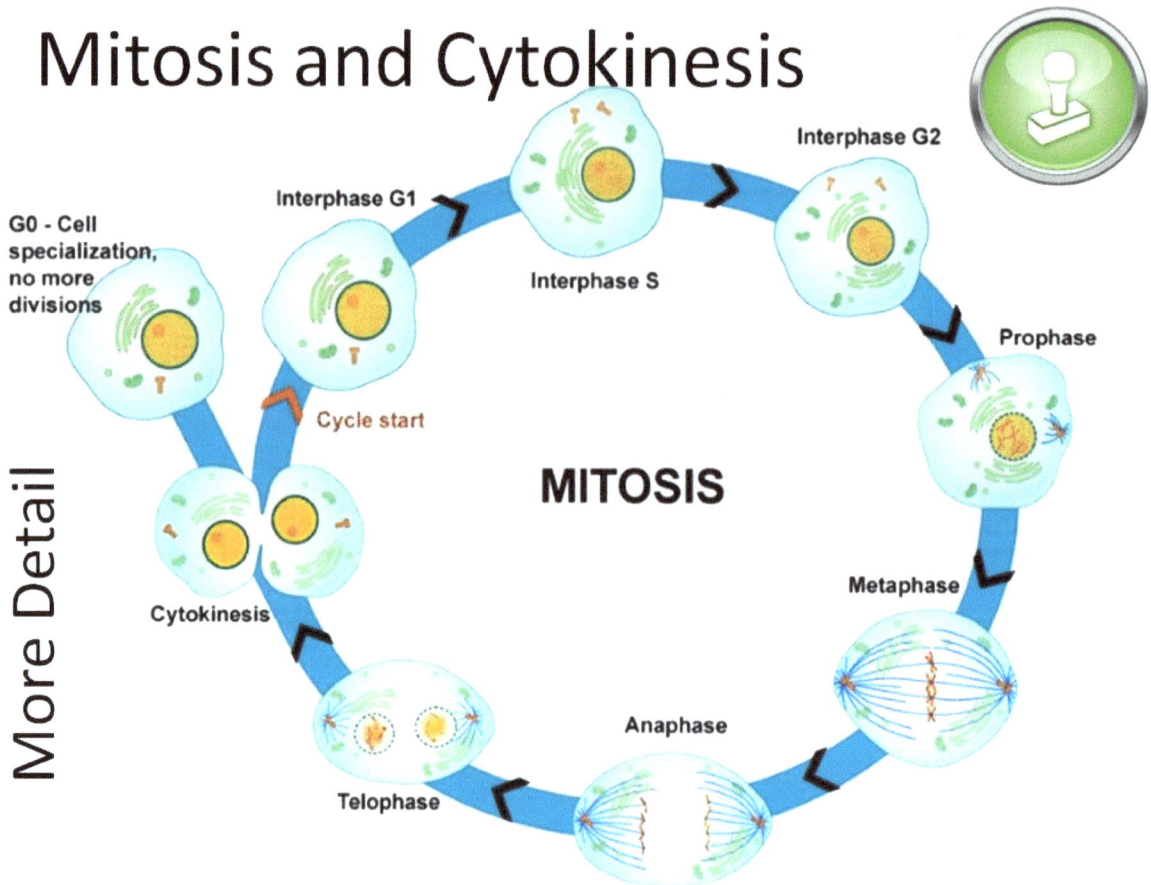

https://www.ck12.org/biology/mitosis/lesson/Mitosis-and-Cytokinesis-BIO/

Draw (Copy) the Illustration Here

18.1 DNA and Heredity

Summarize main points from each video.

Video Title / topic

Video Title / topic

Video Title / topic

Topic Introduction

Summarize your understanding of each paragraph.

DNA stands for Deoxyribonucleic acid. DNA is a molecule. DNA carries genetic instructions for living organisms. DNA is one of the four major types of macromolecules. These four major macromolecules are essential for all known forms of life.

DNA & RNA are nucleic acids. The four major macromolecules are nucleic acid, proteins, lipids and polysaccharides. DNA molecules consist of two biopolymer strands. The strands are coiled around each other. The shape of DNA strands is called a double helix.

DNA stores biological information. The information is replicated when the two strands separate. The two strands of DNA run in opposite directions to each other. The word used to describe this is "antiparallel."

DNA contains the genetic information that allows all modern living things to function, grow and reproduce. It is unclear how long in the 4-billion-year history of life DNA has performed this function. Possibly, the earliest forms of life used RNA as their genetic material.

https://quizlet.com

Read/Summarize Text

1. **Read the passage.**
2. **Underline key expressions in each sentence.**
3. **Re-write each word (or expression) you underlined.**
4. **Summarize the passage.**

Title of Passage.

> DNA, short for deoxyribonucleic acid, is the molecule that contains the genetic code of organisms. This includes animals, plants, protists, archaea and bacteria.
>
> DNA is in each cell in the organism and tells cells what proteins to make. Mostly, these proteins are enzymes. DNA is inherited by children from their parents. This is why children share traits with their parents, such as skin, hair and eye color. The DNA in a person is a combination of the DNA from each of their parents.

https://simple.wikipedia.org/wiki/DNA

Re-write words you underlined

_____   _____   _____

_____   _____   _____

Using a complete sentence, summarize or rephrase the passage

Read Text for Comprehension

Read this article for deeper understanding. No summary is required, although you may want to circle, underline, or mark key ideas and words.

Primary structure consists of a linear sequence of nucleotides that are linked together by phosphodiester bonds. It is this linear sequence of nucleotides that make up the Primary structure of DNA or RNA. Nucleotides consist of 3 components:

Nitrogenous base
- Adenine
- Guanine
- Cytosine
- Thymine (present in DNA only)

5-carbon sugar *which is called deoxyribose (found in DNA) and ribose (found in RNA).*

One or more phosphate groups.

Purine. The nitrogen bases adenine and guanine are purine in structure and form.

Pyrimidines. Cytosine, thymine and uracil are pyrimidines.

For both the purine and pyrimidine bases, the phosphate group forms a bond with the deoxyribose sugar through an ester bond.

A Nucleic acid sequence is the order of nucleotides within a DNA (GACT) is determined by a series of letters.

Sequences can be complementary to another sequence in that the base on each position is complementary as well as in the reverse order.

DNA is double-stranded containing both a sense strand and an antisense strand. Therefore, the complementary sequence will be to the sense strand.

https://en.wikipedia.org/wiki/Nucleic_acid_structure

Draw Illustration

Copy and Label the Illustration in the Space Provided

Illustration

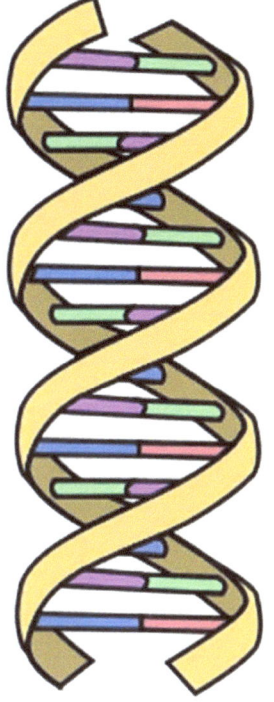

https://frr.wikipedia.org/wiki/DNA

Draw (Copy) the Illustration Here

Interpret a Graph

Write the title of the graph _____

Circle the type of chart this represents
     Bar Chart   Line Chart   Pie Chart   Other

If applicable,
    What does the X-axis represent _____

    What does the Y-axis imply _____

Summarize what this graph represents or conveys

_____

https://en.wikipedia.org/wiki/Genome_size

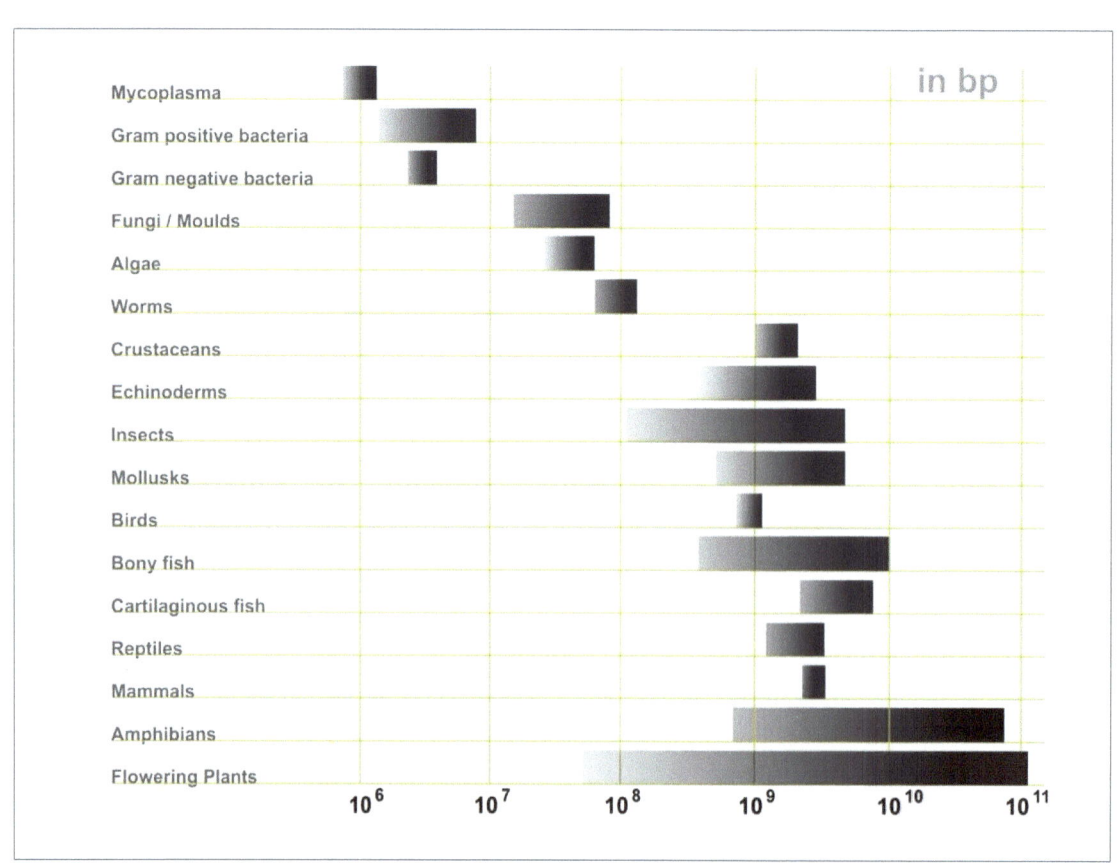

19.1 Genes Genetics and Chromosomes

Summarize main points from each video.

Video Title / topic _____

Video Title / topic _____

Video Title / topic _____

Topic Introduction

Summarize your understanding of each paragraph.

Genetics is the study of genes, genetic variation, and heredity in living organisms. It is generally considered a field of biology. The father of genetics is Gregor Mendel, a late 19th-century scientist and Augustinian friar

Trait inheritance and molecular inheritance mechanisms of genes are still primary principles of genetics in the 21st century. Genetics has given rise to a number of subfields. Organisms studied span the domain of life, including bacteria, plants, animals, and humans.

Genetic processes work in combination with an organism's environment and experiences to influence development and behavior, often referred to as nature versus nurture.

A classic example is two seeds of genetically identical corn, one placed in a temperate climate and one in an arid climate. The one in the arid climate only grows to half the height of the one in the temperate climate due to lack of water and nutrients.

https://en.wikipedia.org/wiki/Genetics

Read/Summarize Text

1. Read the passage.
2. Underline key expressions in each sentence.
3. Re-write each word (or expression) you underlined.
4. Summarize the passage.

Allele

An allele is a variant form of a given gene. Sometimes, different alleles can result in different observable phenotypic traits, such as different pigmentation. A good example of this trait of color variation is the work Gregor Mendel did with the white and purple flower colors in pea plants; discovering that each color was the result of a "pure line" trait which could be used as a control for future experiments. However, most genetic variations result in little or no observable variation.

The word "allele" is a short form of allelomorph which was used in the early days of genetics to describe variant forms of a gene detected as different phenotypes.

https://en.wikipedia.org/wiki/Allele

Re-write words you underlined

_____   _____   _____

_____   _____   _____

Using a complete sentence, summarize or rephrase the passage

51

Read Text for Comprehension

Read this article for deeper understanding. No summary is required, although you may want to circle, underline, or mark key ideas and words.

America's Elite Cows Don't Give Birth — Their Surrogates Do

Embryo transfer, which maximizes the number of offspring a female can reproduce, was first successfully reported in rabbits in 1890. Walter Heape, a pioneer in reproductive biology from England, collected embryos from a black rabbit and inserted them into a white rabbit. A month later, the white rabbit gave birth to a litter of black bunnies. The first ET calf, born in Wisconsin, came along in 1951.

Panda, standing six feet tall and weighing almost a ton, is everything a show cow should be: broad-backed and round-rumped, with sturdy legs holding up her heft. Her hide — thick and black, with splotches of creamy white — fits her name.

"She's a big-time cow," says Dan Byers, owner of Byers Premium Cattle, Inc. "She's a freak of nature is what she is."

Because of her impeccable physique, Panda's descendants sell for a high price. Byers, an elite-cattle breeder in Roseville, Ill., owns several of Panda's daughters. He sold one of her grand calves last year for $10,000 to a family in Oklahoma that shows cattle at state fairs and national competitions.

Cows, like humans, take about nine months to carry a calf to term. At 8 years old, Panda should have seven calves. But in the 1970s, American cattlemen began bucking the reins of nature's limitations by performing a procedure called embryo transfer, or ET, as it's referred to in the industry.

Now, elite-cattle breeders and commercial beef and dairy producers use ET to reproduce dozens of calves a year from their genetically superior heifers, who never actually have to birth a single calf. Surrogates carry the embryos to term.

The process can be time consuming and costly. But Byers, who has bred his genetically superior cattle this way for six years, says it can pay off in the long run.

ET entails several steps. First, the cow's owner injects her with a series of hormones, so she'll produce multiple eggs.

Next, she's bred by a choice stud bull — either the old-fashioned way or, more typically, through artificial insemination. The fertilized eggs, or embryos, are complete genetic packages carrying the bull and dame cow's traits.

http://www.npr.org

Draw Illustration

Copy and Label the Illustration in the Space Provided

Illustration

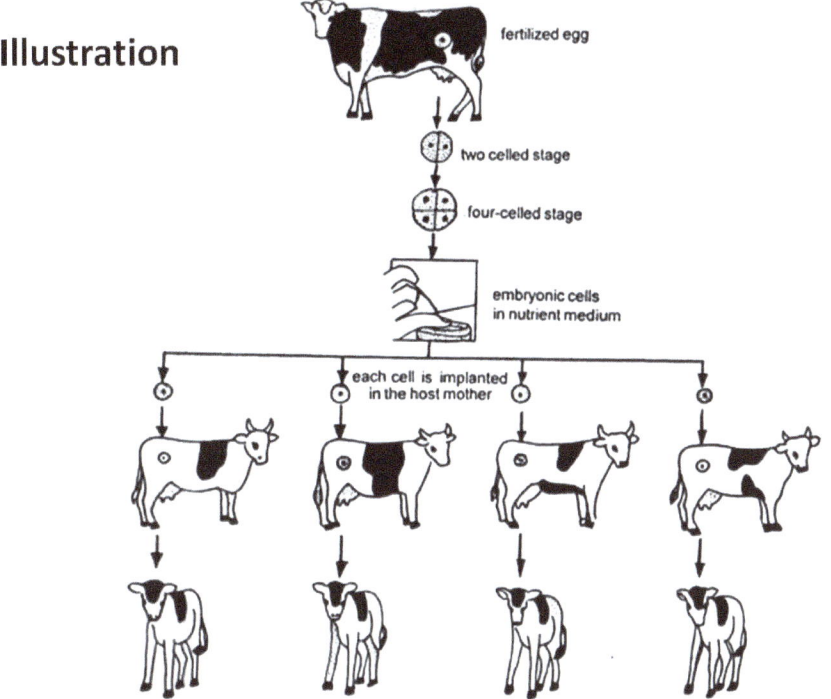

Fig. 8.16. Japanese method of cloning cattle. http://www.biologydiscussion.com/

Draw (Copy) the Illustration Here

Interpret a Graph

Write the title of the graph _____

Circle the type of chart this represents

      Bar Chart   Line Chart   Pie Chart   Other

If applicable,

    What does the X-axis represent _____

    What does the Y-axis imply _____

Summarize what this graph represents or conveys

_____

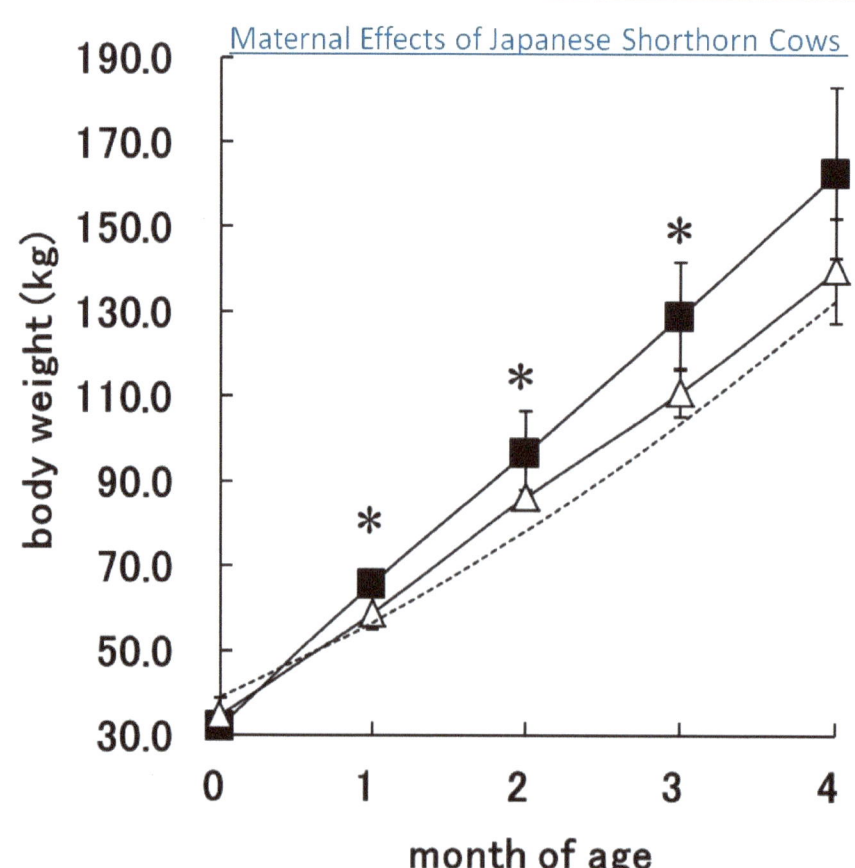

https://www.ajas.info/upload//thumbnails/ajas-26-7-930-5f2.gif

Maternal Effects of Japanese Shorthorn Cows

21.1 Organization of Living Things

Summarize main points from each video.

Video Title / topic

Video Title / topic

Video Title / topic

Topic Introduction

Summarize your understanding of each paragraph.

Scientists often group things into like-kind. This grouping of things results in a "taxonomy" ... taxonomy is the classification of something, especially organisms.

[ ]

In biology class, we will focus on the taxonomy (classification) of living things. But, the expression *taxonomy* also applies to other natural sciences. The expression taxonomy even applies to social sciences. For example, we use "Bloom's Taxonomy" to write learning objectives.

[ ]

Though there is more than one system used to classify objects, science most commonly uses the Linnean system to group organisms.

[ ]

Originally developed by Carolus Linneas in the mid-1700s, this system is used to classify organisms starting with the broad grouping and ending with the most specific.

[ ]

Read/Summarize Text

1. **Read the passage.**
2. **Underline key expressions in each sentence.**
3. **Re-write each word (or expression) you underlined.**
4. **Summarize the passage.**

Classification.

> Classification is the process of organizing different objects into categories based on their common characteristics. Think about the dresser in your bedroom. Typically, one drawer is used for shirts, another for shorts, and yet another for socks. Each clothing item was classified based on how it is worn, and is then grouped with similar clothing in the dresser drawer.
>
> A similar process is used in life science to group organisms. There are numerous different organisms in the universe, each with a unique set of characteristics. To organize them, scientists use a system called taxonomy. Taxonomy is the science of identifying, naming, organizing, and classifying organisms.

cK12.org

Re-write words you underlined

_____   _____   _____

_____   _____   _____

Using a complete sentence, summarize or rephrase the passage

Read Text for Comprehension

Read this article for deeper understanding. No summary is required, although you may want to circle, underline, or mark key ideas and words.

History of Biological Classification

The taxonomic term familia was first used by French botanist Pierre Magnol in his Prodromus historiae generalis plantarum, in quo familiae plantarum per tabulas disponuntur (1689) where he called the seventy-six groups of plants he recognised in his tables families (familiae). The concept of rank at that time was not yet settled, and in the preface to the Prodromus Magnol spoke of uniting his families into larger genera, which is far from how the term is used today.

Carolus Linnaeus used the word familia in his Philosophia botanica (1751) to denote major groups of plants: trees, herbs, ferns, palms, and so on. He used this term only in the morphological section of the book, discussing the vegetative and generative organs of plants. Subsequently, in French botanical publications, from Michel Adanson's Familles naturelles des plantes (1763) and until the end of the 19th century, the word famille was used as a French equivalent of the Latin ordo (or ordo naturalis). In nineteenth-century works such as the Prodromus of Augustin Pyramus de Candolle and the Genera Plantarum of George Bentham and Joseph Dalton Hooker this word ordo was used for what now is given the rank of family.

In zoology, the family as a rank intermediate between order and genus was introduced by Pierre André Latreille in his Précis des caractères génériques des insectes, disposés dans un ordre naturel (1796). He used families (some of them were not named) in some but not in all his orders of "insects" (which then included all arthropods).

https://en.wikipedia.org/wiki/Family_(biology)

Adam Names the Animals

...[19] Out of the ground the LORD God formed every beast of the field and every bird of the sky, and brought them to the man to see what he would call them; and whatever the man called a living creature, that was its name. [20] The man gave names to all the cattle, and to the birds of the sky, and to every beast of the field ...

http://biblehub.com/genesis/2-20.htm

Draw Illustration

Copy and Label the Illustration in the Space Provided

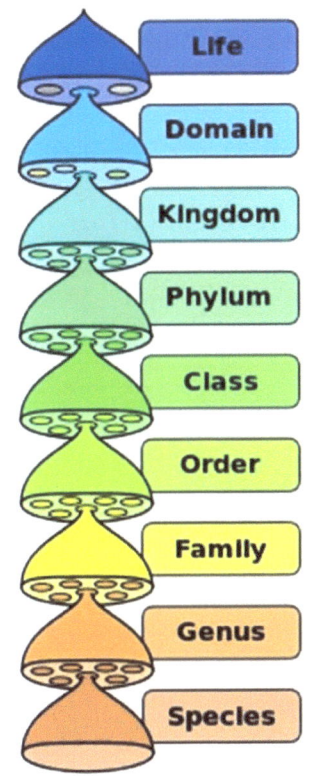

https://en.wikipedia.org/wiki/Family_(biology)

Draw (Copy) the Illustration Here

Interpret a Graph

Write the title of the graph _____

Circle the type of chart this represents

    Bar Chart   Line Chart   Pie Chart   Other

If applicable,

    What does the X-axis represent _____

    What does the Y-axis imply _____

Summarize what this graph represents or conveys

_____

https://commons.wikimedia.org/wiki/File:AnimalsRelativeNumbers.png

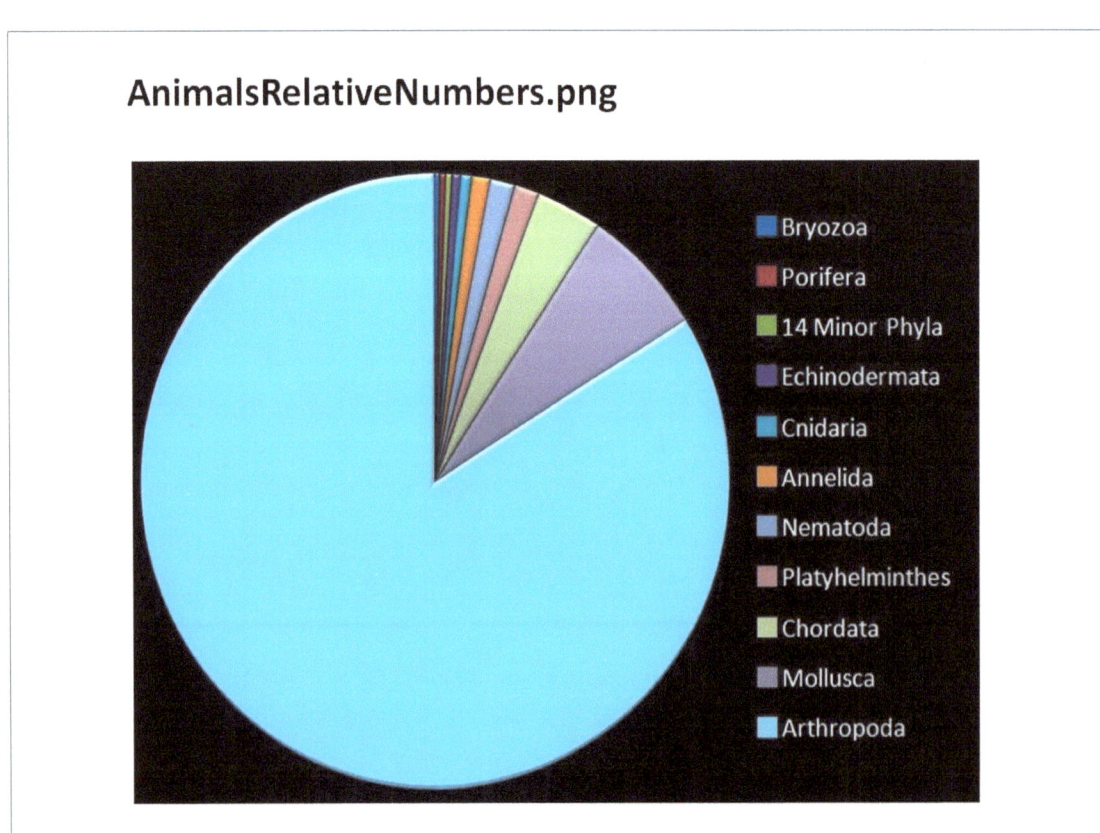

AnimalsRelativeNumbers.png

23.1 Interdependence of Living Things

Summarize main points from each video.

Video Title / topic

Video Title / topic

Video Title / topic

Topic Introduction

Summarize your understanding of each paragraph.

Energy flow, refers to the flow of energy through a food chain. Plants produce their own food using sunlight matter from soil. Herbivores eat plants. Carnivores eat herbivores and sometimes they eat plants or other carnivores. Decomposers break this down – into soil.

In an ecosystem, ecologists try to quantify energy flow. They use math to estimate this. They calculate the relative importance of different component species and feeding relationships.

Solar energy is used by primary producers, like green plants. Primary consumers (herbivores) eat/consume and absorb most of the stored energy in the plant through digestion, and transform it into the form of energy they need

Secondary consumers, carnivores, then consume the primary consumers. Tertiary consumers, consume the secondary consumers. A final link in the food chain are decomposers which break down the organic matter of each of the earlier levels.

Read/Summarize Text

1. **Read the passage.**
2. **Underline key expressions in each sentence.**
3. **Re-write each word (or expression) you underlined.**
4. **Summarize the passage.**

LS2.B: Cycles of Matter and Energy Transfer in Ecosystems.

Plants or algae form the lowest level of the food web. At each link upward in a food web, only a small fraction of the matter consumed at the lower level is transferred upward, to produce growth and release energy in cellular respiration at the higher level. Given this inefficiency, there are generally fewer organisms at higher levels of a food web. Some matter reacts to release energy for life functions, some matter is stored in newly made structures, and much is discarded. The chemical elements that make up the molecules of organisms pass through food webs and into and out of the atmosphere and soil, and they are combined and recombined in different ways. At each link in an ecosystem, matter and energy are conserved.

Oklahoma School Testing Program – Science, Grade 10 2016-2017. (pg 74)

Re-write words you underlined

_____   _____   _____

_____   _____   _____

Using a complete sentence, summarize or rephrase the passage

Read Text for Comprehension

Read this article for deeper understanding. No summary is required, although you may want to circle, underline, or mark key ideas and words.

The energy is passed on from trophic level to trophic level and each time about 90% of the energy is lost, with some being lost as heat into the environment (an effect of respiration) and some being lost as incompletely digested food (egesta). Therefore, primary consumers get about 10% of the energy produced by autotrophs, while secondary consumers get 1% and tertiary consumers get 0.1%. This means the top consumer of a food chain receives the least energy, as a lot of the food chain's energy has been lost between trophic levels. This loss of energy at each level limits typical food chains to only four to six links.

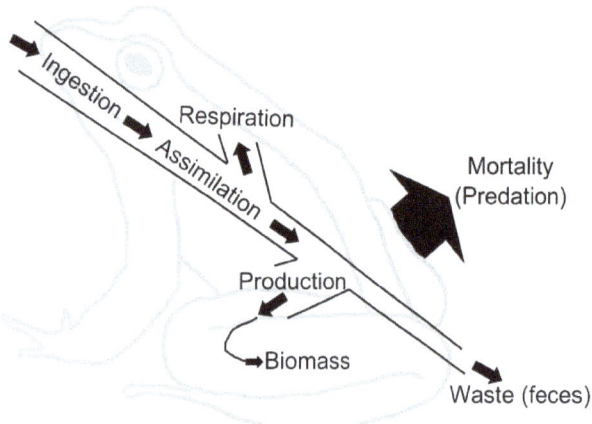

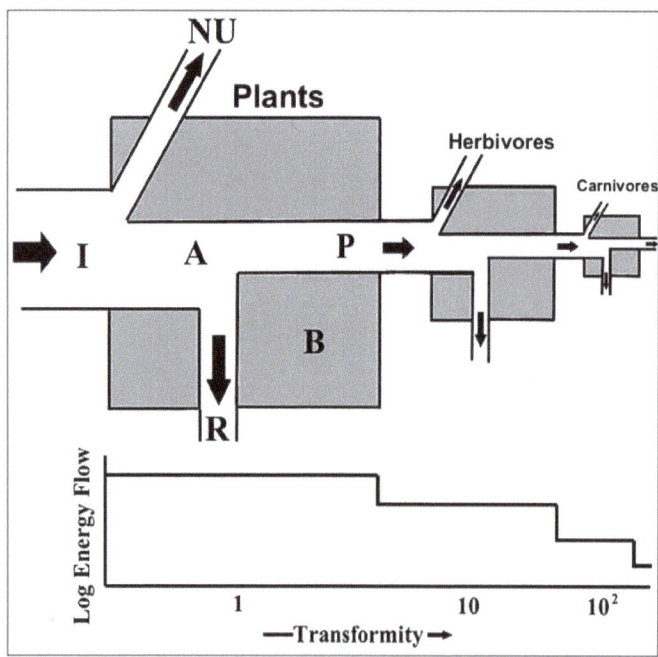

https://en.wikipedia.org/wiki/Energy_flow_(ecology)

Draw Illustration

Copy and Label the Illustrations in the Space Provided

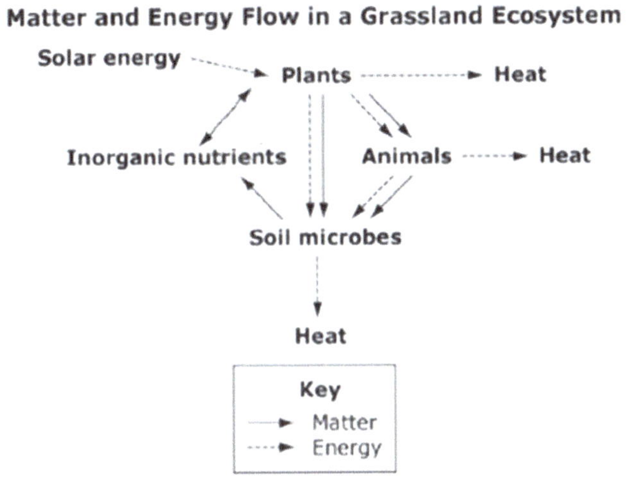

Matter and Energy Flow in a Grassland Ecosystem

Solar energy → Plants → Heat
Inorganic nutrients ↔ Animals → Heat
Soil microbes → Heat

Key
→ Matter
---→ Energy

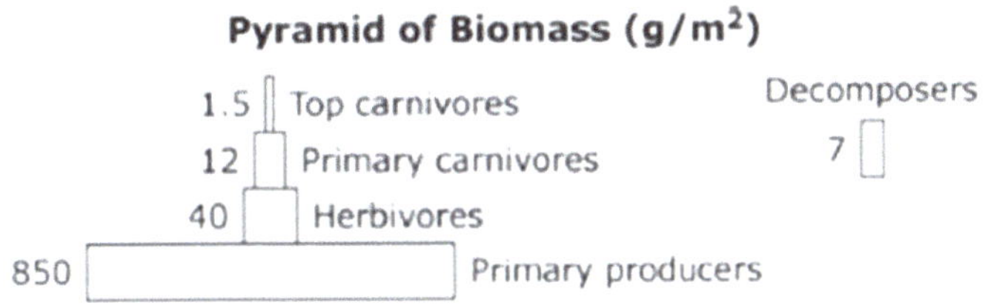

Pyramid of Biomass (g/m²)

1.5 Top carnivores
12 Primary carnivores
40 Herbivores
850 Primary producers
Decomposers 7

Copy the Illustrations – Answer These Questions.

Based on the diagrams, which mathematical expression correctly compares the amounts of energy in different parts of the ecosystem?

A producer energy > herbivore energy
B carnivore energy > herbivore energy
C carnivore energy = herbivore energy
D producer energy = herbivore energy

Based on the diagrams, what is another mathematical expression that correctly compares the amounts of energy in parts of the ecosystem?

A microbe energy = carnivore energy
B herbivore energy > microbe energy
C microbe energy > carnivore energy
D herbivore energy = microbe energy

Use Your Math Skills

A group of students studied a grassland ecosystem. The students learned that biomass is a measure of the amount of matter in an ecosystem. They also learned that energy is primarily transferred through an ecosystem in the form of food. The students created a diagram to show what they learned.

Matter and Energy Flow in a Grassland Ecosystem

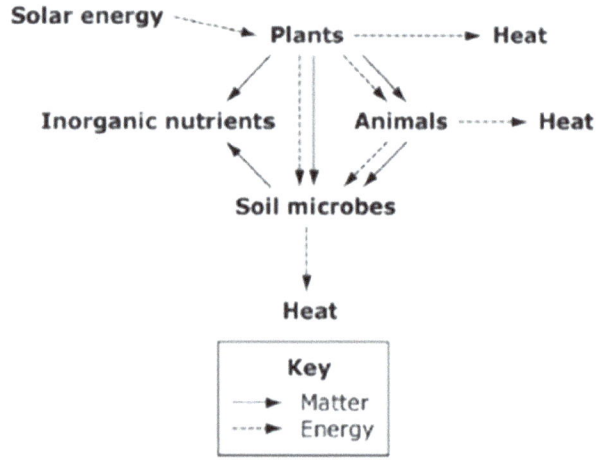

After the students created the diagram, their teacher asked them to answer this question: *How is biomass related to energy flow in the grassland ecosystem?*

To help them answer the question, the students found biomass data. They created this second diagram to illustrate the data.

Pyramid of Biomass (g/m²)

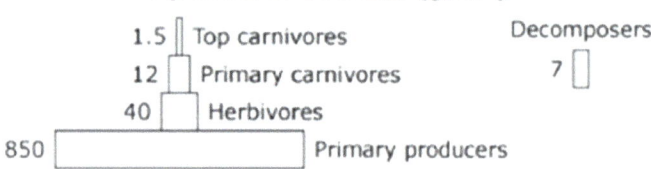

Complete the mathematical expression to compare the amounts of energy in different levels of the ecosystem. Drag and drop the labels into the boxes to create the mathematical expression for the amounts of energy at the different levels. To drag a label, click and hold the label, and then drag it to the desired space. You may use each label once or not at all.

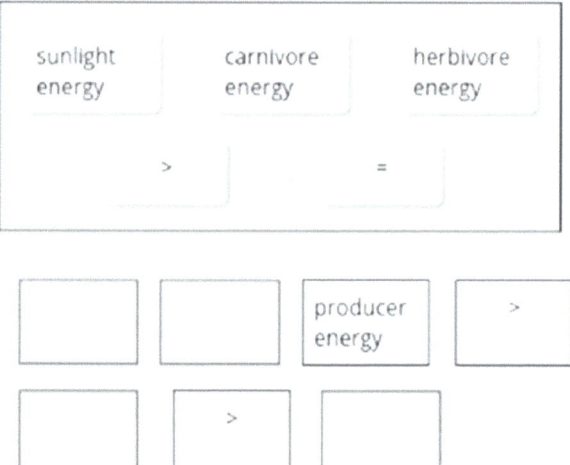

Activity Extract
Item Type: TEI - DOK: 2
Page 175 Oklahoma Science – Grade 10

24.1 Theory of Evolution

Summarize main points from each video.

Video Title / topic

Video Title / topic

Video Title / topic

Topic Introduction

Summarize your understanding of each paragraph.

This topic focuses on the Theory of Evolution. Many associate this topic with Charles Darwin from a voyage in 1831 and his subsequent publication of his observations. Even so, several others began contemplating the subject over 100 years earlier.

Examples of theories that preceded Darwin include catastrophism, gradualism, uniformitarianism, and others. Each of these ideas helped shape Darwin's eventual hypothesis regarding variations in species, adaptation, heritability, and concepts of natural selection.

Darwin's line of thinking – initially as a hypothesis – has withstood the test of time with peer-reviewed/peer-observation and other tests and validation. The expression Theory implies a significant level of testing and rigor has been validated by many scientists.

Aside from direct/visual observations of living things, there is an abundance of other evidence from fossils (paleontology), DNA sequence analysis, embryological evidence, molecular evidence, and protein comparisons across cell-types.

Read/Summarize Text

1. Read the passage.
2. Underline key expressions in each sentence.
3. Re-write each word (or expression) you underlined.
4. Summarize the passage.

Evolution unites all fields of biology.

Scientists are still actively studying evolution through natural selections. The theory of natural selection combined with genetics is sometimes called the modern synthesis of evolutionary theory. The 21st century is an exciting time to study evolutionary biology.

The basic principles of evolution are used in fields such as medicine, geology, geography, chemistry, and ecology.

As much as we know about life on Earth, there is so much more waiting to be discovered.

Excerpts from Biology, Holt/McDougal, page 301.

Re-write words you underlined

_____   _____   _____

_____   _____   _____

Using a complete sentence, summarize or rephrase the passage

Read Text for Comprehension

Read this article for deeper understanding. No summary is required, although you may want to circle, underline, or mark key ideas and words.

Charles Robert Darwin was an English naturalist, geologist and biologist. He is best known for his contributions to the science of evolution.

He established that all species of life have descended over time from common ancestors and, in a joint publication with Alfred Russel Wallace, introduced his scientific theory that this branching pattern of evolution resulted from a process that he called natural selection, in which the struggle for existence has a similar effect to the artificial selection involved in selective breeding.

Darwin published his theory of evolution with compelling evidence in his 1859 book On the Origin of Species, overcoming scientific rejection of earlier concepts of transmutation of species. By the 1870s, the scientific community and much of the general public had accepted evolution as a fact.

Darwin's early interest in nature led him to neglect his medical education at the University of Edinburgh; instead, he helped to investigate marine invertebrates. Studies at the University of Cambridge (Christ's College) encouraged his passion for natural science. His five-year voyage on HMS Beagle established him as an eminent geologist whose observations and theories supported Charles Lyell's uniformitarian ideas, and publication of his journal of the voyage made him famous as a popular author.

Puzzled by the geographical distribution of wildlife and fossils he collected on the voyage, Darwin began detailed investigations and in 1838 conceived his theory of natural selection. Although he discussed his ideas with several naturalists, he needed time for extensive research and his geological work had priority. He was writing up his theory in 1858 when Alfred Russel Wallace sent him an essay that described the same idea, prompting immediate joint publication of both of their theories.

Darwin's work established evolutionary descent with modification as the dominant scientific explanation of diversification in nature. In 1871 he examined human evolution and sexual selection in The Descent of Man, and Selection in Relation to Sex, followed by The Expression of the Emotions in Man and Animals (1872). His research on plants was published in a series of books, and in his final book, The Formation of Vegetable Mould, through the Actions of Worms (1881), he examined earthworms and their effect on soil.

Darwin has been described as one of the most influential figures in human history, and he was honored by burial in Westminster Abbey.

https://en.wikipedia.org/wiki/Charles_Darwin

Draw Illustration

Copy and Label the Illustration in the Space Provided

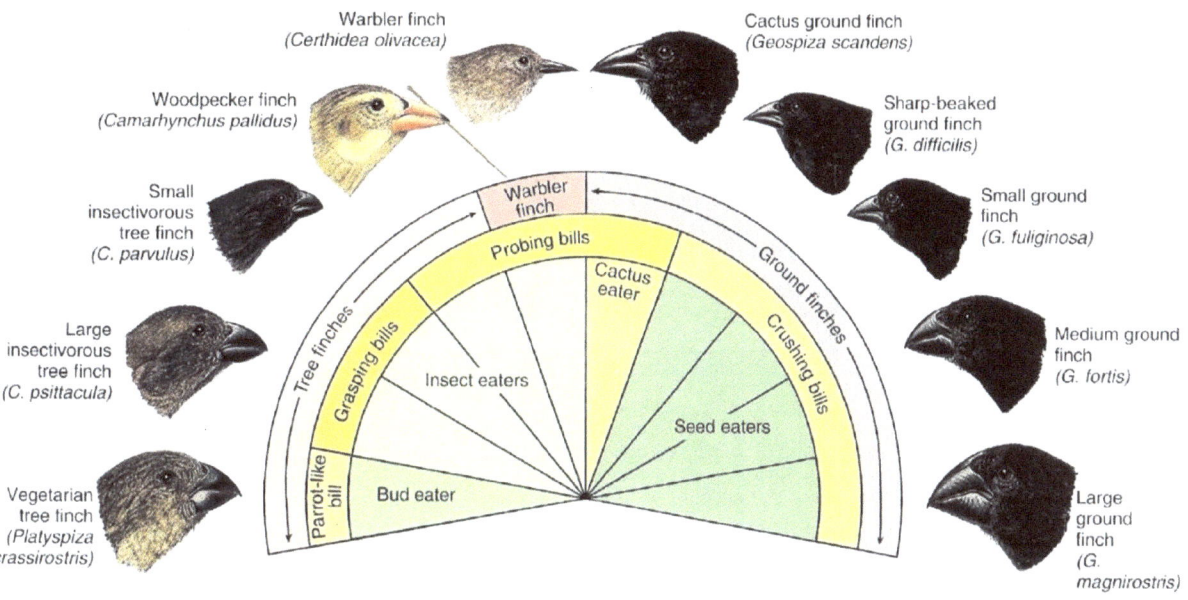

http://www.basfeijen.nl/evolution/pic/finchesbeaktypes.jpg

Interpret an Illustrative Graphic

Describe / interpret the graphic illustration shown below:

_____

_____

_____

By Original uploader was User:TimVickers, SVG conversion by User:User_A1 - Own work (Original text: Self made.), Public Domain, https://commons.wikimedia.org/w/index.php?curid=9381199

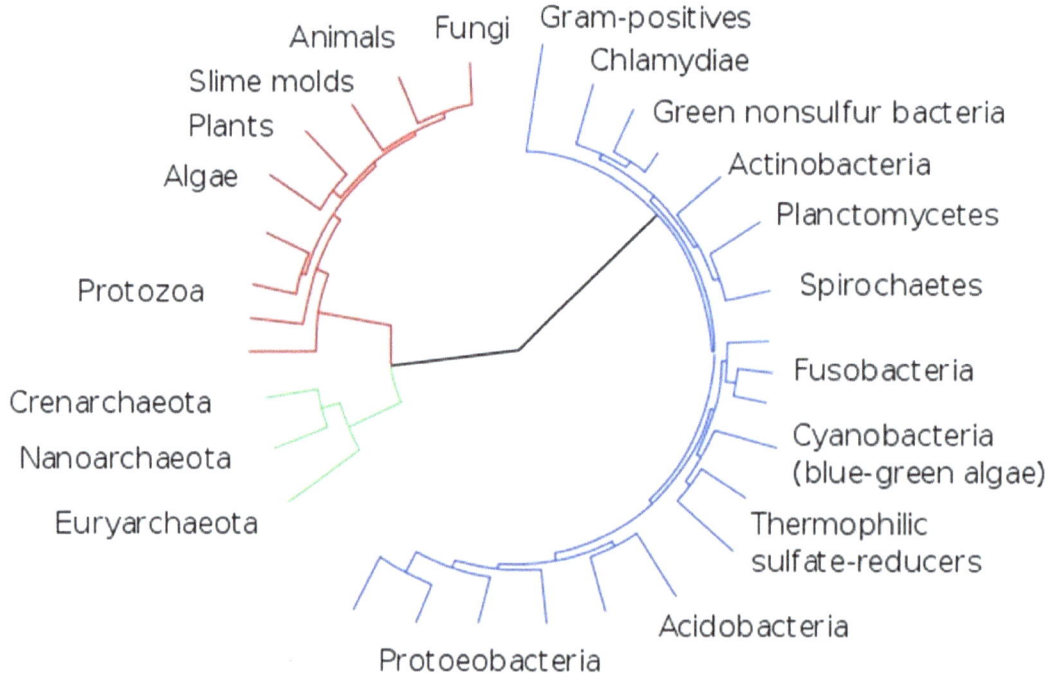

References

Web Sites

acs.org
ajas.info
basfeijin.ni
biologydiscussion.com
ck12.org
crix.org
gc.ca
howstuffworks.com
irank.org
kahnacademy.org

npr.org
quizlet.com
sciencedaily.com
sde.ok.gov
slideplayer.com
time.com
USDA.gov
utah.edu
wikimedia.org
wikipedia.com

Text Books

Allison, M. A. (2010). Austin, TX: Holt McDougal.

Zumdahl, S. S. (2007). Belmont, CA: Brooks/Cole.

Nowiki, S. (2012). Orlando, FL: Houghton Mifflin Harcourt Publishing Company.

Dobson, K. (2008). Austin, TX: Holt, Rinehart and Winston.

For more information, visit

www.HoneycuttScience.com

www.ingramcontent.com/pod-product-compliance
Lightning Source LLC
Chambersburg PA
CBHW051159220526

45473CB00003B/838